CRYPTOCURRENCY, BLOCKCHAINS AND NFT TECHNOLOGY

The very tenets of cryptocurrencies - it's decentralized, democratized, secure from payment fraud, and traceable - render it attractive as a financial alternative to those looking for high returns. In addition to this, it is also a welcome entry after the Great Recession of 2008 that was largely attributed to centralized financial institutions. It is not just the future of financial transactions but is also a marker of how far 21-st century technology has come, in terms of innovation, ease of use, and speed. Even the financial giants such as Signature Bank and JPMorgan Chase believe that cryptocurrency is extremely promising and has the potential to dramatically reduce the expenditure of banks on things like paper checks processing. This has enhanced interest in it, with people keeping a watchful eye over the market.

Dr. Blockchain

CRYPTOCURRENCY, BLOCKCHAINS AND NFT TECHNOLOGY

BY

Dr. Blockchain

TABLE OF CONTENTS

1

What Are Cryptocurrencies?

A cryptocurrency (or crypto currency) is a controversial digital asset designed to work as a medium of exchange that uses cryptography to secure its transactions, to control the creation of additional units, and to verify the transfer of assets. Cryptocurrencies are a form of digital currencies, alternative currencies and virtual currencies. Cryptocurrency also called encryption is a digital or virtual currency that is designed to function as a means of exchange. Use cryptography to secure and verify transactions, as well as organize the creation of new units of a particular cryptotype. Essentially, cryptoscopes are limited database entries that no one can change unless certain conditions are met.

Cryptocurrencies use decentralized control as opposed to centralized electronic money and central banking systems. The decentralized control of each cryptocurrency works through a blockchain, which is a public transaction database, functioning as a distributed ledger. you take away all the noise around cryptocurrencies and reduce it to a simple definition, you find it to be just limited entries in a database no one can change without fulfilling specific conditions. This may seem ordinary, but, believe it or not: this is exactly how you can define a currency.

Take the money on your bank account: What is it more than entries in a database that can only be changed under specific conditions? You can even take physical coins and notes: What are they else than limited entries in a public physical database that can only be changed if you match the condition than you physically own the coins and notes? Money is all about a verified entry in some kind of database of accounts, balances, and transactions.

History Of Cryptocurrencies

In early 2009 an anonymous developer or group of developers introduced it as Satoshi Nakamoto Bitcoin. Satoshi described this as an "electronic framework system of e�ual". It is completely decentralized, which means that no servers are involved and there is no central control body. The concept is very similar to peer-to-peer file sharing networks. One of the main problems that the pay network has to pay is twice the cost. It's a fake techni�ue to spend twice the same amount. The traditional solution was a trusted third party, a central server, which kept records of the situation and transactions. However, this method still included power that actually had control over its assets and all its personal data.

Types Of Cryptocurrency

The first blockchain-based cryptocurrency was Bitcoin, which still remains the most popular and most valuable. Today, there are thousands of alternate cryptocurrencies with various functions and specifications. Some of these are clones or forks of Bitcoin, while others are new currencies that were built from scratch.

Bitcoin was launched in 2009 by an individual or group known by the pseudonym "Satoshi Nakamoto."[1] As of November 2021, there were over 18.8 million bitcoins in circulation with a total market cap of around $1.2 trillion, with the figure updating fre?uently. Only 21 million bitcoins will ever exist, preventing both inflation and manipulation.

Some of the competing cryptocurrencies spawned by Bitcoin's success, known as "altcoins," include Solana, Litecoin, Ethereum, Cardano, and EOS. By November 2021, the aggregate value of all the cryptocurrencies in existence is over $2.4 trillion—Bitcoin currently represents approximately 42% of the total value.[3]

Although cryptocurrency bills itself as a form of money, the Internal Review Service (IRS) considers it a financial asset or property. And, as with most other invcstments, if you reap capital gains in selling or trading it, the government wants a piece of the profits.

2

Where do Cryptocurrencies Get Their Value?

Cryptocurrencies are volatile, and we all are well aware of it. The news about recent crypto highs and lows pops up every day in our feeds, making for the general impression that crypto is a tricky thing to deal with.

Nevertheless, investing in crypto can still be a lucrative investment opportunity, if you know how its value is formed. The following piece sums up common factors affecting the value of digital currencies and indicators, reflective of their truthful value.

So What Is Behind The Crypto Value?

Like any currency, cryptocurrencies gain their value based on the scale of community involvement (like the user demand, scarcity or coin's utility). Still, having in mind, most of the digital coins on the market are issued by private blockchain-related corporations, some factors of crypto value will stem from the image and efficiency of these companies (like project's viability and perceived value). Let us make a general overview of what makes cryptocurrencies valuable.

Coin's Utility

To make a cryptocurrency valuable one needs to make it utile. Any cryptocurrency is primarily a manifestation of using a decentralized digital ledger — blockchain technology. So to make your crypto coin utile, you need to make it usable within a certain blockchain ecosystem.

Let us take Ethereum as a use case. You cannot start using the Ethereum platform without an Ether — a coin, specially tailored to "fuel" the transactions within the Etereum platform. Accordingly, the value of Ethereum depends on the demand for the platform's services.

Cryptocoins' utility can also include dividend payments, mode of exchange within a blockchain ecosystem, voting rights etc.

Scarcity Of The Crypto

Scarcity stands for the finite nature of the digital coins. In the perfect scenario, the demand should excel the supply of the coins, to make it more valuable. For example, the finite supply of Bitcoin never goes beyond 21 million coins. As the most popular crypto in the market, Bitcoin thus enjoys great demand and a rise in value. In a bid to fuel the rise in value, some currencies apply a so-called "burning" mechanism, destroying a part of the coin supply.

Perceived Value Of The Project

Any cryptocurrency value depends on the overall viability and progress of the project development. Projects that keep developing, achieving one milestone after another, establishing lucrative partnerships or launching user-friendly software becomes more valuable in the eyes of the market. All of these are indicators, largely contributing to the positive sentiment around the project and affecting the value of its cryptocurrency.

Why Market Cap Matters More Than The Individual Coin Price?

Market capitalization is a straightforward indicator of the coin's value on the market. The Market cap index is determined by multiplying the total circulating supply by the individual price of the coin.

Market cap = Total Circulating Supply * Price of each coin.

Let us examine a use case. If Coin A has 200,000 coins circulating on the market with each one worth 3$, the market cap of the crypto would be 200, 000*3=$600,000.

In the same way, if Coin B has 100,000 in circulation with each worth $4, the market cap would be 100,000*4= $400,000.

Even though the price of Coin B is individually higher, the total value of Coin A appears much more than Coin B. Thus, the index of the coin market cap is a better way to indicate the true price of a cryptocurrency.

Why use Satoshi pricing in determining the crypto value?

Let's start from the beginning. Satoshi is the creator(s) of Bitcoin (the pseudonym anyway). As a tribute to him/her/them the crypto community named the smallest unit of Bitcoin after him.

So a Satoshi is e□uivalent to 0.00000001 BTC and Satoshi pricing is using this unit as a yardstick - the only 'point of reference' to trade most of the 1500 cryptocurrencies out there.

The perfect analogy for this is USD. You know how the USD is the point of reference to trade not only fiat but all oil and all other commodities? It's the same exact thing. In order to acquire most cryptocurrencies out there is through buying Bitcoin first.

To conclude, Bitcoin and cryptocurrencies are considered volatile with high fluctuations all around. But with an increasing number of tech giants and influencing people showing an interest in blockchain and digital ledgers, and with many governments around the world scratching their heads to find ways to regulate it, cryptocurrency is surely a term that is here to stay and, dare we say, is the future of all currencies.

How Miners Create Coins And Confirm Transactions

Let's have a look at the mechanism ruling the databases of cryptocurrencies. A cryptocurrency like Bitcoin consists of a network of peers. Every peer has a record of the complete history of all transactions and thus of the balance of every account. A transaction is a file that says, "Bob gives X Bitcoin to Alice" and is signed by Bob's private key. It's basic public key cryptography, nothing special at all. After signed, a transaction is broadcasted in the network, sent from one peer to every other peer. This is basic p2p-technology. Nothing special at all, again.

There were many attempts to create a digital currency during a technological boom in the nineties, with systems like Flooz, Beenz and Digi Money that appeared on the market, but inevitably collapsed. The reasons for their failure were many, such as fraud, financial problems, and even friction between company employees and their bosses. In particular, all of these systems used access to trusted third parties, which means that companies behind them checked and facilitated transactions. Due to the shortcomings of these companies, the establishment of a digital billing system is considered a long-lost case. Crypto-currencies are so-called. Because the consensus-building process provides powerful cryptography. This, added to the above factors, makes third parties and blind faith a completely unnecessary concept.

Factors that Determine the Value of Cryptocurrencies

Principles of Free Market Economy (Mainly Supply and Demand)

Supply and demand is a major determinant of the value of anything of value, including cryptocurrencies. This is because if more people are willing to buy a cryptocurrency, and others are willing to sell, the price of that particular cryptocurrency will increase, and vice versa.

Mass Adoption

Mass adoption of any cryptocurrency can shoot its price to the moon. This is due to many cryptocurrencies having their supply capped at a particular limit and, according to economic principles, an increase in demand without a corresponding increase in supply will lead to a price increase of that particular commodity.

Multiple cryptocurrencies have invested more resources to ensure their mass adoption, with some focusing on the applicability of their cryptocurrency to pressing personal life issues, as well as crucial day-to-day cases, with the intention of making them indispensable in everyday life.

Fiat Inflation

If a fiat currency, like the USD or GBP, becomes inflated, its price rises and its purchasing power drops. This will then cause cryptocurrencies (let's use Bitcoin as an example) to increase with respect to that fiat. The result is that you will be able to ac⍰uire more of that fiat with each bitcoin. In fact, this situation has been one of the major reasons for Bitcoin's price increase.

Scams and Cyber Attack History

Scams and hacks are also core factors affecting the value of cryptocurrencies, as they are known to cause wild swings in valuations. In some cases, the team backing a cryptocurrency may be the scammers; they'll pump the price of the cryptocurrency to attract unsuspecting individuals and when their hard-earned money is invested, the price is shortened by the scammers, who then disappear without a trace.

It is therefore imperative to be careful of cryptocurrency scams before investing your money.

Some other factors to consider, that have an impact on the value of cryptocurrencies, include:

- Manner in which the cryptocurrency is stored, as well as its utility, security, ease of ac⍰uisition and cross-border acceptability
- Strength of the community backing the cryptocurrency (this includes funding, innovation and the loyalty of its members)
- Low associated risks of the cryptocurrency as perceived by investors and users
- News sentiment
- Market li⍰uidity and volatility of the cryptocurrency
- Country regulations

Advantages And Disadvantages Of Cryptocurrency

Advantages

Cryptocurrencies hold the promise of making it easier to transfer funds directly between two parties, without the need for a trusted third party like a bank or credit card company. These transfers are instead secured by the use of public keys and private keys and different forms of incentive systems, like Proof of Work or Proof of Stake.6

In modern cryptocurrency systems, a user's "wallet," or account address, has a public key, while the private key is known only to the owner and is used to sign transactions. Fund transfers are completed with minimal processing fees, allowing users to avoid the steep fees charged by banks and financial institutions for wire transfers.

Disadvantages

The semi-anonymous nature of cryptocurrency transactions makes them well-suited for a host of illegal activities, such as money laundering and tax evasion. However, cryptocurrency advocates often highly value their anonymity, citing benefits of privacy like protection for whistleblowers or activists living under repressive governments. Some cryptocurrencies are more private than others.

Bitcoin, for instance, is a relatively poor choice for conducting illegal business online, since the forensic analysis of the Bitcoin blockchain has helped authorities arrest and prosecute criminals. More privacy-oriented coins do exist, however, such as Dash, Monero, or ZCash, which are far more difficult to trace.

3

How To Invest In Cryptocurrency

There are many cryptocurrencies on the market that have different fundamental values. Investors should recognize that a cryptocurrency can be here one day and gone the next, which could leave your investment worthless. That's why it's important to have a strategy around investing in cryptocurrencies and know how to manage your risk.

Cryptocurrency trading beginners may want to consider things like transaction fees, the type of cryptocurrencies available on the platform, special offerings like resources for education and other features that align with your interests and goals.

There are many cryptocurrency exchanges from which to choose. TradeStation, Coinbase, eToro and Gemini, among others, offer an easy, accessible and secure platform to own and transact Bitcoin.

When investing in cryptocurrency, consider the role it will play in your portfolio.

Experts say it's best to take a balanced approach toward investing in cryptocurrencies. Putra says a small portion, between about 2% and 5%, can safely be allocated to crypto in your investment portfolio because the volatility of this asset can cause its value to change dramatically.

Investing in the cryptocurrency market space is often complex especially for the traditional investors. This is because investing directly in Cryptocurrency re⍰uires the use of new technologies, tools and adopting some new concepts.

If you decide to dip your toes in the CryptoCurrency world, you will need to have a clear picture of what to do and what to expect.

Be it Bitcoin, Litecoin, Ethereum or any of the 1300 tokens, buying and selling cryptocurrencies re⍰uire you to choose an Exchange that deals in the products you want.

Being the most famous decentralized cryptocurrency, Bitcoin leads the crypto space so dominantly that the terms crypto and bitcoin are sometimes used interchangeably. However, the matter of fact is that there are other cryptocurrencies as well that can be relied upon for making crypto-investments.

Litecoin

Litecoin, also referred to as 'silver to Bitcoin's gold' is an open-source decentralized payment network that functions without involving an intermediary.

How does Litecoin vary from Bitcoin? Well, both are similar in many ways, however the block generation of Litecoin is much faster than that of Bitcoin. This is making the investors around the world open to accept Litecoin.

Charlie Lee, a former engineer at Google founded Litecoin in 2011. Though Litecoin does not have the anonymity technology of Bitcoin, recent reports have shown that Litecoin is preferred after bitcoin because of its persistence. Another factor that favours Litecoin is the Bitcoin SegWit technology that means secure peer-to-peer trading of currencies without involving exchange participation.

Ethereum

Launched in the year 2015, Ethereum is a decentralized software platform that enables distributed applications and smart contracts to function without third party interference. The currency is the ether that is like an accelerator within the ethereum platform. In the leading cryptocurrencies space, Ethereum. is the second most preferred choice after Bitcoin.

Zcash

Zcash gained attention in the latter part of 2016 and focuses on solving the problem of anonymous transactions. To understand the currency, let's take it as "if bitcoin is like HTTP for money, Zcash is HTTPS".

The currency offers the choice of the shielded transaction to maintain the transparency, privacy and security of transactions. This means the investors can transfer data in the form of encrypted code.

Dash

Originally known as darkcoin, Dash is a more selective version of bitcoin. It was launched in January 2014 by Evan Duffield with the name Xcoin. It is also known as the Decentralized Autonomous Organization or simply DAO. The coin was meant to eradicate all the prevailing limitations of Bitcoin. Currently, Bitcoin has earned a substantial position in the space of cryptocurrencies.

The alternative to virtual currency that promises secured and anonymous transactions through peer-to-peer networking is cryptocurrency. The key to making a lot of money is to make the right investment at the right moment. In comparison to making the everyday money, cryptocurrency models function without involving any middle man as a decentralized digital mechanism. In this distributed cryptocurrency mechanism, the continuous activity is issued, managed and endorsed by the community peer network. The cryptocurrency is known for its speedy transactions over any other mode such as digital wallets and other mediums.

In addition to the above discussed, other top cryptocurrencies include Monero (XMR), Bitcoin Cash (BCH). EOS and Ripple (XRP).

Although bitcoin is the trend setter and is leading the race, other currencies have also made their significant position and are growing in preference each day. Considering the trend, the other cryptos will have a long way to stay and might soon give Bitcoin a real tough time to maintain its position.

What To Consider Before Investing In Cryptocurrency

Investing in cryptocurrencies is very speculative.

Despite stories of investors making millions, investing at an inopportune time can result in rapid and extreme losses.

Although the chance of striking it rich by investing in cryptos is enticing, it's important to understand the characteristics of cryptocurrency. First and foremost, this market is extremely volatile. An asset that can rise so □uickly is also prone to e□ually severe drops.

Another risk: Unlike in other markets, the future of cryptocurrency regulation is uncertain. Some countries that allow the more or less free use of Bitcoin include the U.S., Canada and Australia, to name a few. El Salvador even adopted Bitcoin as a legal tender. But other countries, like South Korea, are pushing restrictive regulation on cryptocurrency, while China has essentially banned cryptocurrency. Even in the U.S., new legislation targets crypto investments for taxation. Cryptocurrencies are conceived of as a unit of exchange, but today, there are only a handful of businesses that accept crypto as a form of payment. Crypto advocates support its utility for broad economic use, but this adoption could take time since regulators around the world are critical of the digital asset.

Calculating the intrinsic value of cryptocurrency may be more difficult than for a publicly traded company, but learning about the asset and how it performs might help prevent you from investing at a peak.

By incorporating industry knowledge and developing an understanding of the digital currency market, you will become a more educated cryptocurrency investor.

Forgery of the cryptocurrency is not possible as the whole system is based on hard core math and cryptographic puzzles. Only those people who are capable of solving these puzzles can make changes to the database which is next to impossible. The transaction once confirmed becomes part of the database or the block chain which cannot be reversed then.

Cryptocurrency is nothing but digital money which is created with the help of coding techni□ue. It is based on peer-to-peer control system. Let us now understand how one can be benefitted by trading in this market.

Cannot be reversed or forged: Though many people can rebut this that the transactions done are irreversible, but the best thing about cryptocurrencies is that once the transaction is confirmed. A new block gets added to the block chain and then the transaction cannot be forged. You become the owner of that block.

Online transactions: This not only makes it suitable for anyone sitting in any part of the world to transact, but it also eases the speed with which transaction gets processed. As compared to real time where you need third parties to come into the picture to buy house or gold or take a loan, You only need a computer and a prospective buyer or seller in case of cryptocurrency. This concept is easy, speedy and filled with the prospects of ROI.

The fee is low per transaction: There is low or no fee taken by the miners during the transactions as this is taken care of by the network.

Accessibility: The concept is so practical that all those people who have access to smartphones and laptops can access the cryptocurrency market and trade in it anytime anywhere. This accessibility makes it even more lucrative. As the ROI is commendable, many countries like Kenya has introduced the M-Pesa system allowing bit coin device which now allows 1 in every three Kenyans to have a bit coin wallet with them.

Despite the market volatility, cryptocurrency have delivered eye-popping long-term returns to investors, rallying as much as 200 percent overtime. It will be a good idea to be in it for the long-term as that guarantees higher return of investments.

As a first time investor, long-term investments in cryptocurrency is the best way to go.

4

The Future Of Blockchain

Blockchain is an emerging technology, so predictions are still mixed about its potential.

In a TechRepublic Research study, 70% of professionals who responded said they hadn't used blockchain. But 64% of said that they expect blockchain to affect their industry in some way, and most predict a positive result.

A recent Trend Insight Report from analyst firm Gartner made the following forecast:

- Through 2022, only 10% of enterprises will achieve any radical transformation by using blockchain
- By 2022, at least one innovative business built on blockchain technology will be worth $10 billion
- By 2026, the business value added by blockchain will grow to just over $360 billion, then by 2030 grow to more than $3.1 trillion

Cybersecurity is one of the most promising areas of projected growth for blockchain technology. An ongoing challenge for businesses of all sizes is data tampering.

Blockchain technology can be used to prevent tampering, keeping data secure and allowing participants to verify a file's authenticity.

Use In Business

Blockchain is, quite simply, a digital, decentralized ledger that keeps a record of all transactions that take place across a peer-to-peer network. The major innovation is that the technology allows market participants to transfer assets across the Internet without the need for a centralized third party. From a business perspective, it's helpful to think of blockchain technology as a type of next-generation business process improvement software.

Collaborative technology, such as blockchain, promises the ability to improve the business processes that occur between companies, radically lowering the "cost of trust." For this reason, it may offer significantly higher returns for each investment dollar spent than most traditional internal investments. Financial institutions are exploring how they could also use blockchain technology to upend everything from clearing and settlement to insurance.

Charity Usage

Over the last few years, a number of charities and foundations have been trialing Bitcoin donations. These include such well-known organizations as the Red Cross, Save the Children, United Way and the Electronic Frontier Foundation. Cryptocurrency donations to charity are clearly on the rise. Fidelity Charitable, which houses the nation's largest donor-advised fund, received $69 mln in cryptocurrency donations in 2017, up from $7 mln received in 2015 and 2016 combined.

Blockchain Properties

Blockchain technology is like the internet in that it has a built-in robustness. By storing blocks of information that are identical across its network, the blockchain cannot:

1. Be controlled by any single entity.
2. Has no single point of failure.

The most obvious risk of an investment in cryptocurrency is the price volatility. As mentioned above, cryptocurrencies trade more like speculative investments than traditional investments. Cryptocurrencies do not pay dividends or coupons and do not generate cash flows, so the investor must weigh the long-term prospects for price appreciation versus the short-term fluctuations in price. Additional risk considerations include potential future competition and regulation. Janet Yellen, former chair of the Federal Reserve and current U.S. Treasury Secretary, said there remain important questions about the legitimacy and stability of cryptocurrency, calling it an extremely inefficient way of conducting transactions. For those who already own cryptocurrency or who are planning to purchase it in some form, there are other risks. Although the blockchain process is largely immune to hacking, individual cryptocurrency wallets and exchanges can be hacked. Hackers can try to penetrate the computer networks that maintain your assets and attempt to steal your private key. One well-known Bitcoin exchange declared bankruptcy in 2014 after hackers stole hundreds of millions of dollars in Bitcoin. That type of risk is not typical when investing in stocks and mutual funds on major U.S. exchanges. (nerdwallet.com) Clearly there are risks, but there are also signs that cryptocurrency is probably not going away. However, the future cryptocurrency landscape may look very different from the way it looks currently, as improvements are made over time to shore up the weaknesses in today's options. Due to the speculative nature of cryptocurrency, we are not recommending it for client portfolios at this time.

5

WHAT ARE NFTS?

Non-fungible token (NFT) is a term used to describe a unique digital asset whose ownership is tracked on a blockchain, such as Ethereum. Assets that can be represented as NFTs range from digital goods, such as items that exist within virtual worlds, to claims on physical assets such as clothing items or real estate. In the coming years, we will see NFTs used to unlock entirely new use cases that are only made possible by crypto.

While the Ethereum ecosystem is where most NFT activity has taken place to date, NFTs can exist on other smart contract platforms too. This is because, at their core, NFTs are just digital abstractions used to represent assets that are one of a kind. Non-fungible token isn't the most intuitive term since we don't commonly refer to the fungibility of objects in the physical world, but this is an important technical distinction when it comes to how an asset is represented on a blockchain. The goal of this book is not to detail every project within the NFT space, but to give a high level overview of what NFTs are, why they are interesting, and showcase some of their potential uses.

NFTs are powerful because, when combined with other financial building blocks on Ethereum, they allow anyone to issue, own, and trade them. This makes interacting with

NFTs significantly more efficient than in traditional platforms. The same reason why cryptocurrency used in payments is more efficient than traditional payments, that it is borderless and significantly easier to transfer, applies to NFTs. For example, if you want to create tradable in-game items as a game developer, then you can instantly have them be tradable through protocols that allow for decentralized exchange of NFTs. You don't have to create your own marketplace or go through the onboarding process of a centralized platform in order to have the items be tradable.

NFT activity can go well past trading and include actions like being able to borrow and lend, support fractional ownership (e.g. NIFTEX), or use it as collateral in taking out a loan (e.g. NFTfi). The possibilities are endless when you have the ability to combine NFTs with DeFi building blocks. For example, the game Aavegotchi combines DeFi and NFT gaming where each Aavegotchi character represents a user's collateral that is deposited within the lending platform Aave, but you can also battle the characters, level them up, and e□uip wearables that change their traits.

NFTs can cover an extremely wide variety of areas given they are simply digital representations of ownership but there has been significant growth within art and gaming in particular. Note that many digital works of art and gaming items are a subset of a larger category of NFT collectibles. There is also the emerging space of social tokens, which sometimes falls into the NFT category or that is closely related to it.

Characteristics of non-fungible tokens

Uni�ue – Deep inside a non-fungible token, metadata describes what makes this asset different from all the rest. This is a permanent, unalterable record that describes what this NFT represents — almost like the certificate of authenticity that you'd get with a rare painting.

Rare – Scarcity is an important ingredient in the recipe that makes NFTs so attractive. While developers have the freedom to generate an infinite supply of certain assets, they also have the power to limit the number of rare, desirable items in existence.

Indivisible – For the most part, NFTs cannot be split into smaller denominations — they can only be bought, sold and held whole. Remember the rules of non-fungibility: you can't purchase 10% of a plane ticket, or collect 50% of a baseball card.

Main benefits of NFTs

There are three key reasons why tokenization has the potential to make things even better:

Ownership – Blockchain technology helps to prove your ownership while making the change of ownership right a much easier process almost instantly.

Transferable – NFTs can be freely traded on specific markets. NFTs can also solve the problem about "walled gardens" in games, that means that an item can be exchanged for items in a different game, even with a completely different publisher.

Authentic – The biggest benefit, fraud prevention, a problem affecting everything from art to tickets, collectibles and even fashion. The blockchain powered NFTs is the perfect weapon against counterfeiting while giving buyers confidence that they'll get what they paid for.

What are the Different Areas where Non-Fungible Tokens can be utilized?

Gaming — Players can use the NFT to trade their digital assets and earn more points or scores by playing video games. The tokens will be authenticated and verifiable to all the participants of a game. Users can utilize their gaming prowess to the fullest extent and generate revenue on a real-time basis. It can be used to buy weapons, vehicles, and characters. This can also establish a well-organized gaming marketplace in the future, where users interested in the virtual world will be able to purchase many games and compete actively.

Artwork — Artists can also make use of NFT and showcase it to their potential customers. The token will contain all the details like the date of origin, previous history of ownership, information about the particular artist, and the assets' total value. Artists can get the actual market value for their effort in the form of peer to peer payments and need not rely on the services of any intermediaries or third parties.

Identity Management — With the need for secure digital storage of data increasing in this technology-driven world, Non-Fungible Tokens (NFT) can contain uni⍰ue information about a specific asset. Any kind of data like medical history, property details, a personal profile, education qualifications, and financial information can be included.

Media and Entertainment — Piracy, copyright infringement, and plagiarism are significant challenges affecting the media and entertainment industry's growth. NFT can use blockchain technology to prevent such fraud, making replicating ideas, and creative work impossible.

Real Estate — The traditional Real estate industry records all the land titles and property registration through cumbersome paperwork. NFT's will digitize the entire process. The property can be tokenized on the blockchain network and be divided into tradable tokens that can be easily exchanged in the leading secondary markets. This removes the role of middlemen like brokers, banks, agents, and lawyers. It also prevents any dispute in ownership of property between any two parties. In Decentraland, a popular virtual marketplace, NFT can be represented as parcels of land in various districts. The users can monetize the property by leasing it out to other co-gamers or utilizing it for advertising.

Non Fungible Tokens (NFT) can also play an active role in event management, digital asset management, licensing, taking care of certificates, maintenance of attractive collectibles, and the sports industry to store the records of teams and athletes.

6

HISTORY OF NFTs

NFTs have actually been around since 2015, but they are now experiencing a resurgence thanks to several factors. First, and perhaps most obviously, is the normalization and excitement of cryptocurrencies and the underlying blockchain frameworks. Beyond the technology itself is the combination of fandom, the economics of royalties, and the laws of scarcity. Consumers all want to get in on the opportunity to own uni�ue digital content and potentially hold them as a type of investment.

Colored coin (2012)

Colored Coins, which were first created on the Bitcoin blockchain for as low as one satoshi (the smallest unit of money), have a lot in common with current NFTs.

It was born out of the desire to issue real-world assets like real estate on a blockchain. Finally, the constraints of Bitcoin's programming language were insufficient to keep the application alive. The notion of Colored Coins, on the other hand, allowed for experimentation and established the basis for future successful NFTs.

CounterParty (2014)

Counterparty is a peer-to-peer financial network built on the Bitcoin blockchain and an open-source, distributed Internet technology. It was one of the first Bitcoin 2.0 sites, allowing users to develop their own trading assets or currencies. It was used for various activities, including meme trading, as shown in the Rare Pepe Directory.

Counterfeiting was almost impossible with tradable valuables on the blockchain, which was not the case with rare real-world items. That was enough to pi▯ue people's interest, given the problem of counterfeiting in the collectibles sector.

Cryptopunks (2017)

On the Ethereum blockchain, CryptoPunks established the world's first marketplace for rare digital art in October 2017. The project's creators displayed 10,000 distinct cartoon characters that anyone with an Ethereum wallet could claim for free. These were immediately claimed, resulting in a thriving secondary market where collectors traded them for higher rates.

Due to the rapid growth of NFTs, Visa also has had to pay for its CryptoPunk ac▯uisition. The NFT that Visa purchased — CryptoPunk #7610 — was sold for 49.50 ETH, approximately $150,000, according to CryptoPunks developer Larva Labs. That's more than double the price of 21.75 ETH paid by a user named "gmoney" less than a month ago.

CryptoKitties (2017)

CryptoKitties, which went viral and raised $12.5 million in funding, signified the mainstreaming of NFT in 2017. CryptoKitty virtual cats can be bred. They have a unique number and a 256-bit genome containing DNA and many features (pattern, mouth shape, hair, eye shape, base color, accent color, and more) that are handed down to their digital feline offspring.

The cartoon collectibles have reached over 5,000 ETH in volume at the height of CryptoKitty buying, breeding, and trading. Founder Cat #18 was sold for 253 Ethereum (about $110,000 at the time of sale). This was eventually surpassed by the $170,000 selling of "Dragon" for 600 ETH in September 2018.

Explosion of NFTs (2021)

The price of a single NFT is continuing to rise. The most recent "Hashmasks" collectibles garnered almost $9 million in sales in a single week. On January 31st, a single Hashmasks digital artwork sold for $130,000, representing a 100,000 percent increase in three days.

On February 9th, Axie set a new record for nine unique plots of digital land sold for 888.25 ETH ($1,500,000) on the blockchain game "Axie Infinity"

Today, commercial pop culture is following suit. "CryptoKicks" Nike's blockchain-based NFT-sneakers, have a patent. Taco Bell, not to be outdone, responded with a gallery of taco art.

Kings of Leon, Grimes, Flume, and Steve Aoki are just a few of the popular musicians that have made NFTs for sale. According to Rolling Stone, NFTs can be a game-changer for singers who want to give songs, albums, virtual merchandise, and even virtual meeting vouchers to "super-fans" as NFTs.

Up-to-date, Pudgy Penguins, a collection of 8,888 randomly produced avatar NFTs, sold out in only 19 minutes and garnered over 30,753 Ethereum, or $97.6 million in total sales volume, is another fascinating endeavor.

The most recent momentum has also been achieved by the Bored Ape Yacht Club. The Bored Apes reached a new milestone in August 2021, when they sold over $1 million worth of Ethereum in their first NFT sale.

However, the record is still held by the Beeple NFT "EVERYDAY: THE FIRST 5000," which sold at a Christie's auction for an astronomical price of $69,346,250.

Some industry pioneers are also heading to a new NFT trend for 2021: "metaverse". For instance, the League of Legends universe was included in an exclusive capsule collection by Louis Vuitton, which included special Prestige skins for Qiyana, the League of Legends Champion.

Gucci and Roblox's partnership is also one of the most recent examples. A virtual Gucci Garden area was open to everyone on Roblox for two weeks at the end of May, coinciding with the launch of the House's Gucci Garden Archetypes — an immersive multimedia experience in Florence, Italy, that examines and celebrates the house's distinctive creative vision. Gucci also released a limited edition of virtual bags in the game, which sold for $4,115 (350.000 Robux, the game's currency), which is higher than the retail price of $3,400.

What's next?

Beyond traditional financial applications, digital collections offer up blockchain technology to whole new possibilities. NFT has the potential to become an essential element of not just the blockchain ecosystem but also the larger economy by representing tangible assets in the digital world.

The number of possible applications is enormous, and many developers, creators, and businesses will probably come up with fresh and exciting ways to improve this promising technology.

NFT myths

Now that we've given an overview of the space, let's talk about misconceptions.

Scarcity alone drives demand

In the early days of the non-fungible token ecosystem, there was a belief that users would care about the provable scarcity of NFTs, and that they'd rush to buy NFTs simply because they were on a blockchain. Instead, we think demand is driven by more traditional forces: utility and provenance. Utility is the obvious one: I'm willing to buy an NFT ticket because it lets me into a conference, I'm more willing to buy a piece of art if I can show it off in a virtual world, and I'm willing to buy an item if it gives me special abilities in a game. The concept of provenance encapsulates the story behind an NFT. Where did it come from? Who's owned it in the past? As the space matures, the stories of interesting NFTs grow more complex and start to meaningfully impact a token's value.

Smart contracts mean assets last forever

There was also the belief that just because a smart contract was deployed, the assets would last forever. This overlooked the fact that there are other entities (websites, mobile apps) that serve as the portal for regular users to interact with these applications. If these portals go down, the assets lose a lot of their worth. Of course, there may be a future where decentralized applications can be deployed in a fully distributed, "unstoppable" way, but for now, we live very much in a hybrid world.

Abstracting the chain away will solve all our problems

In 2018 and 2019, some projects took the approach of "abstracting away the blockchain", hiding all the mechanics of NFTs to users by providing a hosted wallet with username-password authentication. This was an interesting approach because it allowed for the same streamlined onboarding experiences as centralized applications. The problem was that the interoperability with the NFT ecosystem (virtual worlds, wallets, marketplaces) was lost. We've found that projects that plug into the existing NFT ecosystem and perhaps sacrifice some pieces of usability in the short term tend to be more attractive to the current community of early adopters.

7

HOW DO NFTS WORK?

In the recent years, blockchain technology had taken the online trading into the next level. Recently, the hype centered around NFT. According to a report by market tracker DappRadar, NFT trading volume shows 704% increase from the previous uarter. That means there is a massive amount of trading of digital assets happening worldwide within a few month. With this data, it is inarguable to say that NFT displays a promising future in the trading world. As promising as it is, how does an NFT work ?

NFT is an acronym for non-fungible token which refers to digital assets stored in digital ledgers called blockchain These digital assets represent real-world uniue items in digital forms that can be traded online. The following are few examples of NFT sale:

• NFT of the first ever tweet by Jack Dorsey, one of the CEO and founder of Twitter worth $2.9 million

• NFT of 'Nyan Cat' GIF is sold for 300 Ether (cryptocurrency)

• NFT of 'Charlie Bit My Finger' Video is auctioned at $760,999

How does NFT work?

You might be thinking, "What makes this technology possible?"

Using blockchain, it would take an absurd computer to guess and check the hash (the encrypted algorithm) that would gain access to the keys of the token's ownership. Cracking the code would re⍰uire insane computational power, making it impractical to break. And when something is off on the digital ledger, it can be clearly seen as invalid by the network. It is what is known as a peer-to-peer platform. These platforms function without a central authority like a government or bank. You would have to wipe out every computer and node in the network to destroy it. You only need one copy of the blockchain's ledger for all the transactions to be recorded. You might also make it a goal to destroy every node to stop a widely used blockchain network.

NFTs are tokens built on blocks of unbelievably complex algorithms. Instead of a fungible cryptocurrency like Bitcoin, NFTs denote uniqueness and immutability, as mentioned previously. Each NFT is a digital signature written with a cryptographic algorithm, adding another dimension of uniqueness and scarcity. Once set to a specific digital weapon, property, creature, card, artwork, you name it, it is immutable. Immutability means the data can't be changed, forged, or altered. The transaction is encrypted on the blockchain and every single computer, or node, in the network takes note. If you have just node, with the blockchain ledger saved, you'll have all the previously recorded data. Even if your computer spontaneously explodes, the data will be saved on other nodes.

Basically, an NFT is built on a blockchain, a database that stores cryptographed blocks of data then chains them together to form a single source of truth (SSOT) for the data. Every NFT acts as a certificate to the respective digital asset thus determine its authenticity and value.

Specifically, NFTs can represent digital items of any form like digital art, in-game, real-world items even tweets. The ownership of NFTs. NFTs are developed and stored on blockchain particularly Ethereum blockchain (others such as Tezos and Flow). With blockchain acting as a public ledger, NFTs allow the ownership to be verified and traced easily. The ownership of an NFT is managed by uni□ueID and metadata that cannot be duplicated by any other token, hence an NFT only belongs to one owner at a time.

In term of copyright or licensing rights, it is bounded by the smart contract during when the NFTs are created by the original owner. Therefore, depending on NFTs, some will not grant a buyer with exclusive rights to the digital assets.

Therefore, when creators create an NFT, they can

easily prove their ownership.

determine their scarcity on the assets

obtain royalties each time their assets is sold

sell it on any NFT marketplace (i.e opensea.io, Rarible and Foundation)

Pros and Cons of NFTs

Pros

- At least over the past year, NFTs have exploded in value. If that trend continues, there may be more profit potential.
- You can create and sell your own NFTs at virtually no cost. Any proceeds will be pure profit.
- With nearly everything else in the economy going digital, collectibles may be migrating over to digital as well. This might mean the future of NFTs is indeed bright.
- There are already recognized online marketplaces for NFTs. This makes buying, selling, and even creating tokens fairly easy.

Cons

- NFTs are a new and unproven asset class.
- There's no effective way to know, in advance, what any given NFT value might be.
- There's no way to know if NFTs represent the future of collectibles, or it's just a bubble waiting to burst.
- Since anyone can create an NFT, the market may become saturated.
- The recent price spikes make current and future NFT purchases even riskier.
- There may be the possibility of copyright infringement if your image uses part or all of someone else's work. It's even possible your own work could be easily duplicated.

8

WHAT MAKES NFTs VALUABLE?

NFTs have no intrinsic value. Like cryptocurrencies, they're only worth as much as people are willing to pay for them. Unlike a stock, the value of an NFT is not based on shareholder dividends, projected company growth, or annual profit margins.

Some NFTs have sold for millions of dollars, but even those NFTs are not guaranteed to be worth that much. NFTs do not appreciate in value in the same way that real estate does.

NFTs have become popular for selling digital art. Digital artists can sell their artwork online and even receive royalties if the NFT is sold to a new owner. It's unclear whether or not digital art will appreciate any more than a meme or GIF will.

Remember that many NFTs can be rather cheeky. For example, Charmin (the toilet paper manufacturer) released an NFT for "non-fungible toilet paper." For some investors, that may indicate that NFTs are more of a trend than a legitimate, long-term investment.

So, What makes an NFT valuable?

Put simply? The same things that make anything valuable. How rare it is, the □uality of the work, the name/brand involved in it's creation, how easy it is to buy, and the price someone is willing to pay.

The simplest way to explain the value of NFTs is to compare it to a physical product. Let's go with art, because you can buy both of these in physical form and as an NFT.

If you are an art collector, you will pay big bucks for rare pieces of art from well-known artists.

The exact same applies for NFTs. With digital art, you can get your hands on really rare pieces, from some of the most talented artists on the planet.

The difference is, when you buy an NFT, you know for sure that you are the one and only owner of that super rare piece. And, more importantly, you can prove it!

With physical art, this isn't possible. Forgeries are a huge problem in the art world, and although there are checks that can be made, and pieces can be authenticated by experts, it's not a fool-proof system. Experts can be wrong (or possibly in on the con), and you can end up with a fake despite paying a lot of money.

With NFTs, you get proof from the off that you've got the original piece, and while other digital versions might pop up for common viewing, that doesn't change the fact that you've got the original.

So, proof of ownership is one of the major components of why NFTs have value, alongside the rarity of an item, the brand/artist involved, and the price someone is willing to pay.

If you're just coming across NFTs for the first time, and you're like most people, one of the first □uestions that comes to mind is why are these, often highly pixelated, digital images so valuable.

After all, if you were to spend millions of dollars to buy a piece of work done by Picasso, you could hang it in a gallery and charge □uite a bit of money for tickets to simply look at it. But NFTs are different. They're digital files that can essentially be copy and pasted by anyone.

In fact, if you wanted to, you could simply print a copy of the most expensive NFT ever sold and hang it on your wall, and if the owner of the intellectual property behind the token did the same, you couldn't tell the difference.

So, what is it that makes these digital assets so valuable?

NFTs Are Uni□ue

First and foremost, it's important that you understand what the acronym NFT stands for, which is non-fungible token. The non-fungible part of the name points to the uniqueness of the digital asset. These pieces of art simply can't be replaced.

To understand the difference between a fungible asset and a non-fungible asset, all you need to do is look at Bitcoin, Ethereum, or a wide range of other cryptocurrencies, which are all indeed fungible. After all, there's nothing uni⍰ue about a single bitcoin other than its ownership. One could be replaced with another and the owner wouldn't mind because the value is the same no matter which coin he owns.

On the other hand, your dog is a non-fungible asset, albeit one that's likely a very valuable member of your family. After all, I couldn't walk in with another dog that looks pretty much the same, take yours and leave mine, without you being upset about it.

Your dog is uni⍰ue. You know his personality, he knows yours, he gets along with your kids, friends, and family. Shucks, he's part of the family. There's no way to replace him.

Not that NFTs are living, breathing, animals that become part of the family, but they are just as uni⍰ue and irreplaceable as your beloved family pet; hence, making them non-fungible.

Ultimately, uni⍰ueness adds value. After all, value is a concept that's built on supply and demand. When the supply is high and demand is low, value is hard to come by, but with NFTs, the supply count for each uni⍰ue token is one! As a result, if someone wants to buy it, they're not going to be able to unless they pony up the amount of money that the owner of the NFT is willing to accept for it since there's no competitors to lean on when you don't want to pay the price asked.

NFTs Point to the Utility of the Blockchain

Although the technology that makes up the blockchain was invented in 1991 and has been being perfected for decades, it wasn't until very recently that a mass of developers saw the value and decided to start building utility into the equation.

NFTs point to that utility.

You see, NFTs aren't just digital forms of artwork, they can actually be used. Some of the most common ways they're used include:

Games. One of the first games involving NFTs is known as CryptoKitties. In the game, the NFT artworks are digital images of cats with no two looking exactly the same. Those that play the game are able to adopt, breed, and trade CryptoKitties as if they were a mix between a real life pet and a playing card. Today, CryptoKitties are some of the most valuable NFTs on the market.

Tracking. Several corporations have used NFTs as a way to track and verify the source of basic materials and even end products they sell in their stores. Products are essentially tied to NFTs, and when they trade hands, data is added to the ledger. This is overwhelmingly valuable for companies like jewlers that make promises not to use blood diamonds in their jewelry. After all, if the source and distribution of those diamonds is tracked using the blockchain, every step along the way can be verified, ensuring that money is not making its way into the hands of the terrorists that force slave laborers to mine the diamonds.

Ownership Verification. Because each NFT is uni�ue and the owner of that NFT can be verified through the blockchain, these tokens are often tied to limited edition products, attaching an owner to each individual product within the run. For example, say Nike was launching a new style of shoe, but only decided to make 100 of them. The first pair off the line would be the most valuable ever made. As such, if an NFT was attached to that pair, the owner could verify that not only is the pair of shoes one of only 100 ever made, he could verify that it was the first of the 100 made.

Royalties. When developing an NFT, artists are able to add royalties to the blockchain, giving them capabilities they've never had before. After all, in the past, if an artist were to sell one of their works, the sale was final and there were no future revenues that could be made off of that piece of artwork. However, with NFTs royalties are made every time a piece of art trades hands. Imagine the implications this concept carries for the artist. If Picasso would have attached NFTs to his paintings, his family would still be collecting royalties to this day.

As you can see, NFTs aren't just digital images attached to the blockchain. These are pieces of art that can be used in various ways, with each individual use adding value by way of utility.

NFTs Are an Investment In Far More Than Art

OK, for the average joe, you may see some value in the fact that NFTs are uni�ue and that they have �uite a bit of utility. Not to mention, more uses for them are being developed each and every day. However, that doesn't yet point to hundreds of thousands or even millions of dollars in value.

What I find to be the most valuable part of the NFT is that these digital artworks are investments in much more than the artwork itself or even the utility it represents. They're an investment in the underlying cryptocurrency, known as Ethereum.

Think of it this way:

When you purchase an NFT, you'll purchase it using the cryptocurrency that lives on the same blockchain as it, with the most popular being the ethereum blockchain, which uses ether cryptocurrency.

So, say you wanted to purchase a Cryptopunk for one ether, which in United States dollars works out to about $3,487 as of September 2021. You like the art and you decide to hold it for a while, but in five years, you decide it's time to sell and look for other artworks for your portfolio.

Even if you decided to sell the art for exactly what you paid for it, one ether, there's a strong chance that you'll make a hefty profit when that ether is converted back into USD. After all, five years ago, ether had a value of about $11.

9

HOW TO INVEST IN NFTs?

While cryptocurrencies like Ether, Bitcoin, and even Blockchains have been around for some time, NFTs are a relatively new investment option. They've been around since 2014 but are making a tear in the investment markets now because they are becoming a popular way to buy and sell digital artwork.

While most investors are considering NFTs an alternative investment, is it the right investment for you?

NFTs are attractive because they reassure via storing the ownership data of a digital file on the Ethereum blockchain that the NFT is the only 'authentic' one of its kind. NFTs cannot be altered or counterfeited, so they are considered a secure method of asset ownership.

Yet, alternative investments typically represent a greater risk than mainstream investments because they're highly volatile. If you consider NFTs as possible alternative investments, please do so with just about the amount of money you are willing to lose. Consider that, like most assets, NFTs are unpredictable and have yet to stand the test of time.

NFTs have proven to be a profitable form of investment owing to some of the following reasons:

Creates Value for the Tokenized Asset: NFTs create a medium whereby physical objects like art works can be tokenized, thus eliminating the duplication of such art work and limiting ownership to the artist. This in turn creates scarcity for the art work and hence, value for it.

It provides Investors more Li□uidity: Tokenizing assets gives investors more li□uidity over their assets when they need it. An example is when a virtual land owner decides to rent out his/her virtual space to advertisers or influencers for a fee, while still retaining ownership over the land. The virtual land in this case still belongs to the owner, but part of it is li□uified as rent.

Potential for growth and development: NFTs possess potential for growth and development of the land sector. Pegging NFTs to land pieces has proven great potential for growth and development, for instance in real estate, owning and controlling virtual lands gives you the power to decide what you want to do on your land. You can decide to rent it out, build up a solid and secure business for advertising or online sales. NFTs have proven to be a profitable form of investment owing to some of the following reasons:

Creates Value for the Tokenized Asset: NFTs create a medium whereby physical objects like art works can be tokenized, thus eliminating the duplication of such art work and limiting ownership to the artist. This in turn creates scarcity for the art work and hence, value for it.

It provides Investors more Li□uidity: Tokenizing assets gives investors more liquidity over their assets when they need it. An example is when a virtual land owner decides to rent out his/her virtual space to advertisers or influencers for a fee, while still retaining ownership over the land. The virtual land in this case still belongs to the owner, but part of it is li□uified as rent.

Potential for growth and development: NFTs possess potential for growth and development of the land sector. Pegging NFTs to land pieces has proven great potential for growth and development, for instance in real estate, owning and controlling virtual lands gives you the power to decide what you want to do on your land. You can decide to rent it out, build up a solid and secure business for advertising or online sales.

Steps to Invest in an NFT:

1. **Open an NFT Marketplace Account**

You can browse an online NFT marketplace without signing up for the service. You'll need to create a digital wallet, which will be funded with cryptocurrency. Adding the digital wallet to the online marketplace will open your account. Once you have that up and running, you can participate in the marketplace and invest.

2. Create a Digital Wallet to Buy NFTs

A digital wallet acts like a physical wallet, in that it holds your money, but a digital wallet is specifically designed to hold cryptocurrency.

Generally speaking, the best choice is a digital wallet that uses a thumb drive or other physical media to hold your cryptocurrency, also known as a cold wallet. Since it won't sit actively on the web, it's less likely the wallet will be compromised.

You'll need to get a digital wallet that works with the NFT marketplace you want to invest from (for example, Open Sea works with Ethereum). Your wallet must be compatible with the cryptocurrency you want to buy and sell on the platform. MetaMask supports Ether, for example.

3. Fund Your Account

You must buy a cryptocurrency like Ether to participate in an NFT marketplace.

You can do so quickly and easily through investment brokers that accommodate cryptocurrency trading, such as Webull and SoFi Active Invest. Each provides commission-free trading in individual stocks, exchange-traded funds (ETFs) and options. But they also enable you to purchase popular cryptocurrencies like Ether and Bitcoin.

If you want a dedicated cryptocurrency platform, check out Gemini. It's a full-service cryptocurrency exchange, where you can buy, sell, and store your cryptos. And as a crypto-specific site, Gemini also offers plenty of tools and research information to help you better understand the crypto market.

Once you purchase your cryptocurrency, you can load it onto your digital wallet to use on an NFT marketplace.

4. Buy Your NFT

With your digital wallet active and funded with cryptocurrency, you'll be ready to buy.

It's important to understand NFT marketplaces use auction formats. You'll need to submit a bid for the token you want to purchase. The sale will go through if you are either the highest bidder or the only bidder.

How to Create Your Own NFTs

Create nftTo create your own NFT you'll need to work with a marketplace. We'll use Open Sea as an example, which works with the Ethereum network. If you follow the steps above, you'll already have Ether and the digital wallet to hold it. You'll need both to create an account on Open Sea.

Once you do, you can click "Create" on the top bar of the page, then "My Collections." Under "Create new collection," click "Create," and enter your collection's name with a description and upload an image. Once that's done, hit "Add New Items." It doesn't matter that the image isn't your desired NFT; you're adding it just to get started on the page.

Once you've completed that step, you can make modifications. For example, you can upload metadata like an image, video or audio file. From there, you can rework your image until it's the final product you want for your NFT. You'll need to add a uni⍰ue name to your NFT, as well as a se⍰uence number if you're making multiple versions of the same token.

Once you're satisfied with what you've produced, click "Create" and sign the message in your digital wallet. There are no costs for creating NFTs on Open Sea, but there is a 2.5% fee on the final sale price.

There are various methods for listing and selling your NFT, which will be provided by the digital marketplace. Connecting your NFT availability to your social media pages will likely help you sell it.

What are NFT Marketplaces?

NFT is a (uni⍰ue) digital asset or, an artwork, or a product essentially having a digital certification that can direct upwards to digital asset ownership. Additionally, unlike fungible tokens, it is not possible to buy or sell NFTs directly. It requires a dedicated platform called the marketplace.

Buyers can always browse through marketplaces like OpenSea and look for buying virtual assets using cryptocurrencies (such as Ethereum and Bitcoin). Marketplaces charge a fixed percentage on the total transaction amount related to the ac⍰uired NFT.

How Does an NFT Marketplace Work?

NFT platform user flow:

NFT runs on smart contracts: When a Non-fungible token development company launches a token, it includes a uni�ue piece of code recorded in blockchain for providing immutability.

Usually, there is a fixed user flow of all NFT marketplaces:

Step 1: User signs up/registers their profile on the NFT marketplace platform.

Step 2: The user then needs to set up a crypto wallet.

Step 3: Now, it is the first NFT creation step. The user is supposed to post a digital collectible (any artwork/asset that they aspire to tokenize) to pitch/showcase their work.

Step 4: Now, while listing the sale ad for the asset, a user can choose between asking for a certain amount, or auctioning the asset and ultimately selecting a bid winner as the next owner of NFT.

Step 5: Assets come over the moderation.

Step 6: Once the fresh NFTs get approved, platform users can browse for them in the sales listings.

Step 7: The NFT poster can start accepting and analyzing bids.

Step 8: Once the auction gets over, a notification is sent by the platform to token owners about the highest value bids.

Step 9: The NFT Marketplace platform verifies and avails the transfer of NFT ownership of the digital asset.

Once you become the rightful owner of any NFT product, you can keep checking its prices on the platform. If the prices rise, you can resell them and gain profit out of them. There are so many NFT platforms like OpenSea, MakersPlace, etc. You can invest in one to profit from this rising crypto market or even make a platform if you want!

What Are The Risks?

Investing in NFTs presents awesome opportunities for you, but you also need to be cognizant of a number of risks. Here are some of them:

The Risk of Fraud

Any market, especially on the blockchain, comes with some level of risk. Non-fungible tokens are no different. Although efforts to make blockchain technology more secure, especially the authentication process, cybercriminals, and hackers may still be able to replicate NFTs. If they do, you can buy fake NFTs that are of no value. To avoid this risk, it is paramount to be extra careful and always check the authenticity of any NFT of choice before making payment. You might also want to deal only with trusted sellers on the blockchain.

Smart Contracts

Smart contracts are designed to automate contractual agreements and eliminate third parties. It is considered one of the most advanced inventions in blockchain applications. For example, most decentralized platforms are using smart contracts to offer crypto loans and ensure both parties stick to the agreed terms and conditions. However, it has emerged that hackers can still exploit smart contracts to defraud unsuspecting clients.

In one of the most renowned protocols, the DeFi protocol Poly Network, cybercriminals struck and made away with $600 million worth of NFTs.

Evaluation Challenges

This is another notable challenge when investing in non-fungible tokens. Today, the valuation of NFTs largely depends on creativity, authenticity, and scarcity of owners and buyers. Lack of a standard way of evaluating the tokens means you are at risk of buying overpriced NFTs. Again, maintaining the value of NFTs is pretty challenging, which implies that the risk of selling them at a lower value in the future always looms. If the value drops, perhaps because of shifting demand, it means one will incur a loss.

Market Risks

The digital world is evolving rather fast, and what is selling like hot cake today, such as non-fungible tokens, might be overtaken by other types of assets. Therefore, it is pretty difficult to predict how NFTs will perform in the future. It is crucial to factor in the market and financial risks before investing in NFTs. However, these risks are not native to NFTs only. Even other trading instruments, from ETFs to forex, are still faced with the same challenge.

The best way to address the market and financial risks is by diversifying your investment portfolio. For example, instead of only investing in NFTs directly, it will be wise to use eToro to diversify into NFT stocks and commodities. Again, you need to follow the market carefully to understand when the trend starts changing.

Legal Risks

When you decide to invest in NFTs, it is also essential to appreciate the legal challenges. For example, there is still no clear, well, let's say, official definition of NFTs. This is why different countries have varying approaches when it comes to classifying NFTs. For example, despite the chairman of SEC indicating that NFTs are sold as securities, the Supreme Court ruled that they are investment contracts. This implies that it will take time to get the tokens to comply with the Howey test before they can be considered securities.

Why Are NFTs the Future of Digital Trading?

NFTs are blowing up and NFT development services are much sought after at the moment. If that is not a stable enough indicator that NFTs are the future of digital trading, then take a look at these remarkable characteristics of NFTs that make them poised to take over the world of trading in the coming years:

Authenticity

An NFT gives a valid certificate of authenticity which grants the users legitimacy and ownership rights for any and all items they purchase from an NFT marketplace.

Uni�ueness

No NFT can be interchangeable under any conditions and are one-of-a-kind. This means that the data associated with NFTs cannot be changed or tampered with in any way and it proves the certification for the authenticity of the digital assets.

Scarcity

NFTs create an air of digital scarcity as they are all uni�uely non-interchangeable and re�uire large amounts of energy to create, hence it is not sustainable to create NFTs without a limit.

Li�uidity

NFTs listed in one marketplace are also available for other similar marketplaces, which implies liquidity sharing business concepts.

Transparency

NFTs are developed in a decentralized ecosystem, much like cryptocurrencies, and the transactional data are verified publicly. Hence, NFT provides a high-level of transparency that makes them safe from fraudulence.

Is It a Good Idea to Invest in NFTs?

If you are planning on investing in the future, then yes, it is a good idea to invest in NFTs. NFTs are also closely linked to cryptocurrencies, since they are built on the same blockchain as Etheruem (a globally popular cryptocurrency). This means that they offer all the benefits of cryptocurrencies such as li⍰uidity, absolute transparency, and hinders fraudulence by keeping a public ledger that records all transactions and is accessible to anyone who wants to track or monitor trading activities.

10

Ways To Make Money With NFTs

NFTs can be used by decentralized applications (DApps) to issue unique digital items and crypto-collectibles. These tokens can either be a collectible item, an investment product, or something else.

Gaming economies are nothing new. And since many online games have already had their own economies, using blockchain to tokenize gaming assets is taking only a step further. In fact, the use of NFTs could potentially solve or mitigate the common problem of inflation that many games have.

While virtual worlds are already flourishing, another exciting use of NFTs is the tokenization of real-world assets. These NFTs can represent fractions of real-world assets that can be stored and traded as tokens on a blockchain. This could introduce some well-needed li�uidity to many markets that otherwise wouldn't have much, such as fine art, real estate, rare collectible items, and many more.

Digital identity is also a sector that can benefit from the properties of NFTs. Storing identification and ownership data on the blockchain would increase privacy and data integrity for many people around the world. At the same time, easy and trustless transfers of these assets could reduce friction in the global economy.

If you have been wondering how people make money with NFTs, you'll be glad to know there are several ways to do that. Look at these four ways that have had some tangible results.

Digital Works of Art

Individual NFT artworks are among the most popular and valuable non-fungible tokens created so far in terms of profitability. The history of blockchain and art changed dramatically on March 11, 2021, when an auction house sold artwork for a staggering $69 million. This was the auction house's first time auctioning a fully digital artwork.

NFTs, give the right of ownership to the artist such that when you create a digital asset, you're able to show proof of authenticity and profit from your work. If you happen to be a meme creator, you get to benefit when it circulates. Looking at it from that perspective, NFTs are a revolutionary technology for art creators.

Fashion NFTs

In a similar way as with artwork, NFTs address authenticity issues and eliminate the possibility of counterfeits. Clothing and fashion brands are increasingly benefiting from the NFT trend by releasing a digitized limited edition of their outfits with a collection of iconic celebrity pieces or design signatures. In addition, the clothing and fashion brands can benefit from NFT development by building an NFT marketplace for their clothes and accessories. This allows them to generate unlimited revenue from the NFTs in a secure and robust environment.

Licensed Collectibles

Tokenizing collectibles is among the best niche application of NFTs and feels like it's the most obvious way to deal with collectibles. People already in the sector dealing with physical collectibles such as trinkets, memorabilia, trading cards, and other items now only have to sell the same as digital assets. With NFT's capability to prove uncommonness, the value of the collectibles can be significantly higher than their physical counterparts.

The most popular collectibles so far are sports cards. Initially, the sports card project only allowed the trading of licensed footballer cards. Now, the National Basketball Association(NBA) has also launched its collection of NFT cards. There's a high likelihood that other sports will follow suit, and collections will have a wider variety of sport NFTs to buy.

This is not to say that other collectibles aren't doing well. As stated earlier, physical collectibles can be turned into tokenized assets and bought, traded, or sold as such. While physical trading and handling can damage the collectibles, NFTs never lose their □uality as they only exist in digital form and are securely stored on the blockchain.

Video Games

Video game NFTs are also taking shape as blockchain allows you to turn the games from paying to win to playing to earn. For the games to engage larger audiences and increase revenue opportunities with high-demand rewards, NFT game development is the way to go. The games have functions that allow buying, selling, trading, or exchange of your collectibles. Although so far there aren't any games that have gained high popularity, the potential is enormous.

With games being well-known huge spenders on virtual items, video games with NFTs sellable in-game items have a likelihood of becoming a big winner. Notably, NFT video game developers are also ambitious creators of NFTs, which might thrust the NFT technology further forward.

Compared with other NFTs such as artwork, fashion, and trading cards, in-game NFTs are the most advanced of all. This is because there are complex, fully interactive yet keep changing with time by having to be upgraded or to be leveled with the changes in a player's character.

What are the Latest NFT's Emerging in the Market?

TRON has come out with an NFT named TRC721 on its blockchain network. All the information regarding the Non Fungible Tokens (NFT) are stored safely in the smart contracts. It can be seamlessly integrated with various DApps and DeFi projects of

TRON. This will strengthen TRON's position in the crypto economy and it can use its speedy network and low transaction cost to compete with Ethereum in the NFT space. The TRC721 NFT can be used to represent any digital or tangible asset easily.

Binance is making efforts to launch an NFT incentive program on the Binance Smart Chain. It will be used to empower NGOs and charities in China. Volunteers can buy the NFT and get attractive gifts and rewards. It also offered NFT's for Halloween and Christmas. The Binance Launchpad also offered users to buy digital land NFTs through its platform.

DIGITALAX, an NFT, has been launched on the Ethereum blockchain network. It focuses on the fashion industry. They are working with several fashion designers now globally. The designers get royalties whenever someone purchases their product. DIGITALAX also organizes various auctions for the users to participate on its platform. It also aims to integrate its offering with gaming and virtual reality platforms.

Beeple, an NFT, raised a whopping $3.5 million in an auction conducted by an online auction house named Nifty Gateway. Beeple was created by a graphic designer called Mike Winklemann. Mike believes that digital artists will be able to make more money in the future if there is a change in the current business model.

The Main Obstacles to the Growth of NFT's

The regulatory uncertainty that refuses to make a transition from the rigidity of centralization to the flexibility offered by decentralization.

The lack of ade�uate technical and economic infrastructure affecting the usage of Non-Fungible Tokens (NFT).

Doubts regarding the scalability of blockchain technology are hindering the adoption of NFT's.

Chances of higher transaction fees in case there are some congestion issues in the blockchain network.

Issues in user experience, especially when Non-Fungible Tokens are distributed to a lot of people at the same time.

Top NFT Marketplaces

2021 saw the non-fungible token (NFT) market explode, with some crypto art and other virtual collectibles selling for millions.

In this guide, you will learn about the top seven NFT marketplaces that allow you to buy, sell, and collect NFTs.

OpenSea

OpenSea is a New York-based non-fungible token (NFT) marketplace founded in 2017. On this platform, users can buy and sell blockchain-backed collectibles, art, music, domain names, trading cards, and virtual worlds, among other items.

The OpenSea website is simple and easy to navigate. Users can access all the NFTs in the marketplace, the statistics of the top NFTs on the platform, and the help center for guidance. Beginners will find it straightforward to mint and sell NFTs on OpenSea. Users also enjoy almost gas-free listings after listing their first NFT.

NFT creators on OpenSea have the advantage of earning royalties every time a buyer ac⍰uires their item. You can set up a royalty fee of up to ten percent. Moreover, users can add items from other NFT marketplaces like Rarible and Mintable.

Users on OpenSea can add other networks besides Ethereum to their wallets using Chainlist. Some of the top-performing non-fungible tokens on this marketplace at the time of writing were CryptoPunks, Sorare, and Ethlings.

Rarible

Rarible is an LA-based NFT marketplace that categorizes its NFTs into art, photography, music, domains, DeFi, games, and Metaverses. Here you can buy, sell, and create non-fungible tokens. The marketplace was founded in 2020.

When you visit Rarible for the first time, you will notice that it is ⍰uite busy. Therefore, it might take you a minute to figure out how to get started. You will also see the top-selling NFTs and the hot collections on the homepage. This gives visitors a glimpse of the marketplace activity. Rarible also displays several celebrity NFTs. The royalty fee is ten percent.

The platform provides verified badges to authentic and dedicated collectors and creators. Users have to fill a form to get verified. Additionally, you can view and manage your Rarible collectibles on OpenSea.

You can also list NFTs on OpenSea in RARI. This is the Rarible governance token. Active users that influence the decision-making process of the marketplace receive rewards in RARI. Rarible wants to eventually evolve into a fully decentralized autonomous organization (DAO) where users make all the decisions.

SuperRare

SuperRare is a Delaware-based Ethereum-based marketplace for rare pieces of digital art. Artists showcase their work on this platform by invite-only because SuperRare only wants to display high-□uality work. Moreover, it wants to have a proportional number of active collectors and artists.

SuperRare uses a simple design allowing users to find what they want easily. The platform also promotes new art releases as a way of supporting its artists. SuperRare has built-in royalties for its artists on secondary sales. The primary and secondary sale commissions for creators are 85 and ten percent, respectively. Buyers pay a three percent fee on all purchases.

SuperRare was established in 2018.

Mintable

Mintable is a Singapore-based NFT marketplace where you can mint, buy, and sell non-fungible tokens. The platform was founded in 2018 and supports Ethereum and Zilli□a-based NFTs. Creators can set royalties of ten percent.

Mintable offers several categories of NFTs, including art, game items, collectibles, and music. This marketplace also displays the top NFTs on the homepage. You will also find it effortless to navigate through the website.

There are three types of items on Mintable: gasless, normal, and printable. When an item sells, the platform charges 2.5, five, and ten percent fees on normal, gasless, and printable items. The marketplace also has an NFT DAO allowing community-driven governance. Each voter gets a voting non-fungible token that contains their voting power.

Users can access extra features like a custom store and an advertising package by upgrading to the Mintable pro plan.

Nifty Gateway

Nifty Gateway is one of the top NFT marketplaces. It is head□uartered in New York and was founded in 2018.

Nifty is a centralized marketplace for digital art. Interestingly, you do not need to pay in ETH to use this platform. That is because you can buy and sell nifties (the term for NFTs on the platform) in USD. Regulated US digital currency exchange Gemini owns Nifty Gateway.

The platform releases collections from top artists through drops. These collections are only accessible for a limited period. On the website, you can view the timer showing when the next drop will take place. Also, it is easy to navigate through the platform.

Artists get to decide the royalty fee they will take on secondary sales. Nifty takes a five percent fee plus 30 cents from each secondary sale.

Foundation

Foundation is a digital art platform where the community invites artists. After joining the marketplace, invited artists can then mint and sell their NFTs. Creators receive a commission of 85 percent on primary sales and ten percent on secondary sales. Creators can also invite other creators.

The website design is minimalistic and professional. However, there is a lot of content on the homepage, which could increase the website loading time.

According to Foundation, creators have earned more than $40 million since it launched in February 2021. The famous Nyan Cat NFT sold on this marketplace for 300 ETH ($842,325).

MakersPlace

MakersPlace is an NFT marketplace for digital creators like artists, writers, and photographers. It brings together talented creators enabling them to sell their creations.

The platform offers escrow services and the option to accept credit card payments for secondary sales. Furthermore, the marketplace releases new artwork through drops, just like Nifty Gateway. MakersPlace adds creators through invites and takes a 15 percent fee on all ETH purchases. The royalty fee is ten percent.

MakersPlace has its head□uarters in San Francisco and was founded in 2018.

The listed NFT marketplaces have made NFTs easily accessible to anyone with an internet connection and an Ethereum wallet.

CONCLUSION

The very tenets of cryptocurrencies - it's decentralized, democratized, secure from payment fraud, and traceable - render it attractive as a financial alternative to those looking for high returns. In addition to this, it is also a welcome entry after the Great Recession of 2008 that was largely attributed to centralized financial institutions.

It is not just the future of financial transactions but is also a marker of how far 21-st century technology has come, in terms of innovation, ease of use, and speed. Even the financial giants such as Signature Bank and JPMorgan Chase believe that cryptocurrency is extremely promising and has the potential to dramatically reduce the expenditure of banks on things like paper checks processing. This has enhanced interest in it, with people keeping a watchful eye over the market.

With lawmakers around the world gearing up to draft concrete laws for cryptocurrency regulation, it is important to note that such shifts will have a significant impact on their prices. However, the introduction of Bitcoin ATMs, its adoption by FinTech companies such as PayPal and S□uare, and the vocal backing of CEOs of conglomerates naturally point to a future when goods and services will be increasingly bought with cryptocurrencies.

Volatility is common in the global financial market, and crypto assets are no different. With it slowly entering the mainstream, it will soon mold itself to the ebbs and flows that the market is known for. In the coming years, we will certainly see rapid adoption of cryptocurrencies, with some market pundits going as far as to say that digital currencies will become as ubiquitous as the credit card.

What does this mean for investors who wish to grow their portfolios? The crypto market is certainly an attractive one, with great possibilities and potential. While it makes sense to diversify your portfolio, a wise route would be to steer clear of big bets and also invest in stable markets, keeping in mind the volatility of cryptocurrencies.

Bitcoin has been breaking into all-time highs every month since October 2020. It surpassed the mark of $50,000 at a point, and it seemed to be heading near $60,000 then.

Although it would have been ideal for an investor to have entered the investment space earlier, it does not mean that it is now too late to invest in cryptocurrency, provided that you invest what you can afford.

There may be dips in its value from time to time, but even a small investment you make in cryptocurrency today could return manifold 200percent gains year after year.

Made in the USA
Middletown, DE
05 November 2026

36546283R10044